STEM Women Study Guide

Editor: Deborah J. Levine

Books by Deborah J. Levine

Matrix Model Management System:

Guide to Cross Cultural Wisdom

The Liberator's Daughter

Religious Diversity at Work

Religious Diversity in our Schools

Inspire your Inner Global Leader

Going Southern: The No-Mess Guide to Success in the South

Teaching Curious Christians about Judaism

STEM Women Study Guide

© Deborah J. Levine 2015

Published by American Diversity Report

Chattanooga, TN 37412

1 (888) 451-2798

web: www.americandiversityreport.com

e-mail: info@americandiversityreport.com

ACKNOWLEDGEMENTS

STEM Women Study Guide is a Project of Women Ground Breakers.

Thanking our 2015 Sponsors

Platinum Sponsors:
Chattanooga Area Chamber of Commerce, Humanities Tennessee

Gold Sponsors:
American Diversity Report, Chattanooga Writers Guild, EPB Fiber Optics, excellerate!, Million Women Mentors, Nashville Area Hispanic Chamber of Commerce Foundation, Southern Adventist University, The HR Shop, ThreeTwelve Creative, UTC College of Engineering and Computer Science, Volkswagen Chattanooga.

Special Thanks to Southern Adventist University Intern Abigail White

TABLE of CONTENTS

Bios & Discussion Questions

APPENDIX: 2015 Groundbreaking Storytellers

A. Moderator Sheila Boyington: National Sr. Advisor to STEMconnector®

B. Alyssa Montague: Document Control Manager of Hutton Construction, Inc.

C. Heidi Hefferlin: President at Hefferlin + Kronenberg Architects, PLLC.

D. Jemila Morson: Owner of MorSocial Media.

E. Lakweshia Ewing: Co-owner of Biz Boom Apps, LLC.

F. Dr. Neslihan Alp: Interim Dean at UTC College of Engineering & Computer Science.

G. Sonya Reid: Systems Management Engineer at Erlanger Health System.

H. Tamra Van Allen: Assistant Vice President and Pricing Actuary at Unum.

I. Dr. Ruth Williams: PsychoBiology Professor at Southern Adventist University.

SOURCES - References

ADA LOVELACE (1815-1852)

Ada Lovelace was an English mathematician and the writer of the first published computer program. She was originally named Augusta Ada Byron and was the daughter of the famous poet, Lord Byron, and his wife, Annabella. In 1835, Ada married William King, ten years her senior, and when King inherited a noble title in 1838, they became the Earl and Countess of Lovelace. Most women in her position at that time were not encouraged in their education or intellect. Known as "the first programmer", Ada was assisted in her learned by a mathematician-logician, Augustus De Morgan, who taught Mathematics at the University of London.

While working for an English mathematician, Charles Babbage, Ada developed an interest in his machines which later proved to be the forerunners of the modern computer. In 1843, Ada succeeded in translating and annotating an article written by mathematician Luigi Federico Menabrea on one of Babbage's machines. Using what she called, "Poetical Science", Ada also made detailed description of how an "Analytical Machine" could be programmed to calculate a sequence of rational numbers. Babbage referred to Ada as an "enchantress of numbers." Today the Ada computer programming language developed in the 1980s for the U.S. Department of Defense is named in her honor.

DISCUSSION QUESTIONS:

1. What do you think about combining mathematics and creativity to develop technology and computer programing?

2. What about Ada's story is most memorable to you?

ALICE AUGUSTA BALL (1892-1916)

Alice Ball was an American chemist who invented a chemical extraction process called the Ball Method. She was born in Seattle Washington and is the granddaughter of a slavery abolitionist, J.P. Ball. Alice graduated from the University of Washington in 1912 with a pharmaceutical chemistry degree and a bachelor's degree in 1914. She went on to complete a master's degree, during which she researched how to extract active ingredients from the root of the Kava plant, now used for its sedative and tranquilizing qualities.

After hearing about her skills as a chemist, Alice was asked by a U.S. public health officer to use her technique with chaulmoogra oil, oil that had been associated with the treatment of leprosy for years. Alice succeeded in isolating certain active agents from the oil that later were injected to treat leprosy (also known as Hansen's Disease). This method was used until the 1940s when sulfones came into use for the treatment of leprosy.

Alice was the first woman to graduate from the college of Hawaii with a master's degree in chemistry. After her death at the young age of twenty-four, the Board of Regents of the University of Hawaii awarded her their highest honor, the Regents Medal of Distinction.

DISCUSSION QUESTIONS:

1. What about Alice's story do you find most inspiring?

2. If you were a chemist, what illnesses would you like to research and help treat?

ANITA BORG (1949-2003)

Anita Borg Naffz was a computer scientist who supported the progress of women in the field of technology. Anita studied at University of Washington and went on to receive her doctorate from the Courant Institute at New York University. After receiving her degree, she worked for computer companies such as the Digital Equipment Corporation and Xerox PARC (Xerox Corporation Palo Alto Research Center).

Anita was active in advocating for women in technology. She established an organization called Systers, an online community for women in the field of computing, and the Institute for Women in Technology (later renamed Anita Borg Institute), an organization that supports women's involvement in technology. She was also cofounder of the Grace Hopper Celebration of Women in Computing, which was a technical conference honoring and promoting the work of women in the fields of science and technology. Anita was inducted into the Women in Technology international Hall of Fame in 1998. She was appointed by President Clinton to the Commission on the Advancement of Women and Minorities in STEM, a commission working to strengthen the presence of women and minorities in STEM fields.

DISCUSSION QUESTIONS:

1. What are the activities in your school and community that assist and inspire girls interested in STEM?

2. What are some fun ways that you would like to get involved in a STEM field?

ANNIE EASLEY (1933-2011)

Annie Easley was a programmer who created computer programs that help conserve energy. Annie was born in Birmingham, Alabama and attended Xavier University in New Orleans, Louisiana, where she majored in Pharmacy. Annie married and relocated before completing her degree. Deciding to test out a different career path, she applied for a job with National Advisory Committee for Aeronautics (the precursor organization for NASA) and was hired to work at the Lewis Research Center.

After beginning her new job, Annie's interests took a different direction and she decided to pursue mathematics. Annie completed her B.S. in mathematics from Cleveland State University while working for the NACA. She did research on how to determine the life usage of storage batteries like those found in electronic cars. During her time spent working for NACA, Annie developed computer code used in various energy projects involving solar and wind energy.

DISCUSSION QUESTIONS:

1. What part of environmental science appeals to you as a career path?

2. What do you think you need to learn now to pursue that career?

ASIMA CHATTERJEE (1917-2006)

Asima Chatterjee was a scientist from India who made significant contributions to the field of chemistry. Asima received an Master's degree in science from Calcutta University in 1938 and her Doctorate in 1944. She was the founder and director of the Chemistry Department at the Lady Brabourne College in Calcutta and became an honorary lecturer in chemistry at Calcutta University. Asima later became the Khaira professor of chemistry at Calcutta University, one of the most honored positions at the university.

One of Asima's greatest passions was developing a holistic approach to healing. She researched plants native to India that could be used as ayurvedic drugs for medical purposes. Asima established a Regional Research Institute and an ayurvedic hospital to further holistic medicine. Asima published much of her research including around 400 papers and several review articles. She also was chief editor for a publication by the Counsel for Scientific and Industrial Research entitled *Treatise of Indian Medical Plants.* Asima was a dedicated scientist and an inspiring teacher. She was elected a Fellow of the Indian National Science Academy, was the first woman to receive the Shanti Swarup Bhatnagar Award (a science award given in India), and was appointed to the Rajya Sabha (upper house of India's parliament) by the president of India.

DISCUSSION QUESTIONS:

1. What do you think of holistic medicine as a career path?

2. What professional training / education might you need to work or conduct research in this area?

<u>BESSIE BLOUNT (1914-2009)</u>

Bessie Virginia Blount is an important pioneer in the fields of technology and science. Bessie was born in Hickory, Virginia and attended Panzar College of Physical Education and Union Junior College. She became a physical therapist and worked with World War II veterans who had become amputees. Bessie invented an electrically powered feeder device that could help disabled people feed themselves. She patented the invention in 195 under her married name, Bessie Blount Griffin.

She discussed her inventions with Thomas Edison, and although they were quite revolutionary, she had difficulty cultivating an interest in her inventions in the United States. However, Bessie was recognized by the inventor community globally. In 1969, Bessie's interests changed and she became involved working for law enforcement as a forensic scientist. She was the first African American woman to go to work for Scotland Yard.

In her later years, Bessie launched her own business examining historical documents from before the civil war. She was a handwriting expert, having been punished for being left-handed as a child, a common situation in those times. She forced herself to learn to write with her right hand and even her feet. Her expertise, her perseverance, and her compassion fueled her extraordinary life.

DISCUSSION QUESTIONS:

1. What do you think about the impact of Bessie's left-handedness on her career?

2. Why do you think there was a lack of interest in her inventions in the U.S.?

CAROLYN DENNING (1927-   )

Carolyn Denning is a physician who made great progress working with a disease called cystic fibrosis. As a child, Carolyn did not think she could become a doctor because she did not know that women could become physicians. However, her father felt that she should become a physician, following his recommendation, Carolyn decided to attend medical school. She graduated from Tulane University School of Medicine in New Orleans, Louisiana in 1952.

Carolyn's education included a research fellowship studying cystic fibrosis. She continued to study cystic fibrosis throughout her career, researching, writing, and lecturing about the disease. Carolyn was the first person to use an interdisciplinary approach to the treatment of cystic fibrosis and after forty years working with cystic fibrosis, she helped to bring about significant changes. For example, in 1958 people born with cystic fibrosis usually died in childhood. Now about thirty-nine percent of people with cystic fibrosis live to be eighteen or older.

DISCUSSION QUESTIONS:

1. What is your dream career?

2. What are some ways in which you can use your dream career to help change people's lives for better like Carolyn did?

CHARLOTTE SCOTT (1858-1931)

Charlotte Scott was a mathematician who was born in England to Caleb Scott and Eliza Exley. Charlotte was tutored at home and showed an interest and ability in mathematics at a young age. She earned a scholarship to Girton College at Cambridge in 1876 and later became the first woman to earn first honors class in the Mathematical Tripos Examination (although she was not publicly recognized due to her gender).

Charlotte's work was featured in several publications and she published many works of her own, including her well known book, *An Introductory Account of Certain Modern Ideas and Methods in Plane Geometry* (published in 1894). She was the head of the mathematics department at Bryn Mawr College in Pennsylvania for thirty-nine years and was key in establishing the College Entrance Examination Board, for which she served in the position of chief examiner. Charlotte was the second European woman to earn a doctorate, the first British woman to earn a doctorate, and after moving to the U.S., the first woman in the United States to hold a doctorate in mathematics. She was also the first woman on the American Mathematical Society Council and served as its vice president.

DISCUSSION QUESTIONS:

1. What do you think it was like to be the first woman in so many milestones for the field of mathematics?

2. What kind of math do you like? Dislike?

EMILY WARREN ROEBLING (1843-1903)

Emily Warren Roebling was a builder and a business woman who served as a stand-in engineer for one of the greatest engineering projects of her century, the Brooklyn Bridge. As a girl, Emily's brother Gouverneur Kemble Warren, ensured that she received a secondary education at Georgetown Visitation Convent in Washington D.C. where she earned highest honors.

In 1865, Emily married Washington Roebling, the son of engineer John A Roebling. Washington became involved in a building project with his father constructing what was originally called the Great East River Bridge between Brooklyn and Manhattan. Emily's father in law, who was the chief engineer on the project, died after a surveying accident, leaving Washington as the chief engineer. Several years later, Washington collapsed due health complications related to the construction process. During her husband's illness, Emily carried on the work of building the bridge, conducting negotiations to secure contracts for the building materials needed, recording instructions for the field engineers, and working with the bridge board of trustees.

While she received little recognition during the construction process, Emily was publicly recognized at the dedication ceremony in 1883, was presented to Queen Victoria, and was invited as a guest to the coronation of Emperor Nicholas II of Russia. Always learning and striving, Emily enrolled in law school in 1899 and obtained her law degree before her death at age sixty.

DISCUSSION QUESTIONS:

1. Had you been in Emily's position do you think you would have stepped into the position of leadership or waited see if someone who might be more qualified would take charge?

2. What do you think of Emily's life-long commitment to learning?

EXCERCISE #1
STEM Thinking

1. What does Figure 1 mean to you?
2. What does Figure 2 meant to you?

Figure 1: Engineering Design

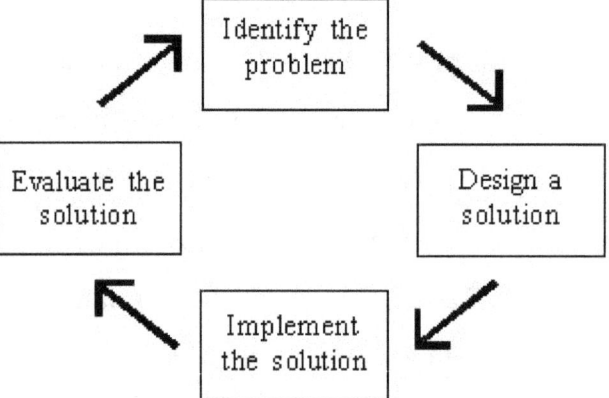

This cycle repeats itself thus making the process dynamic, flexible and creative. The solution that is designed needs to be effective and efficient. Thus it has to achieve its objective without incurring high costs.

Figure 2: Instructional Systems Design

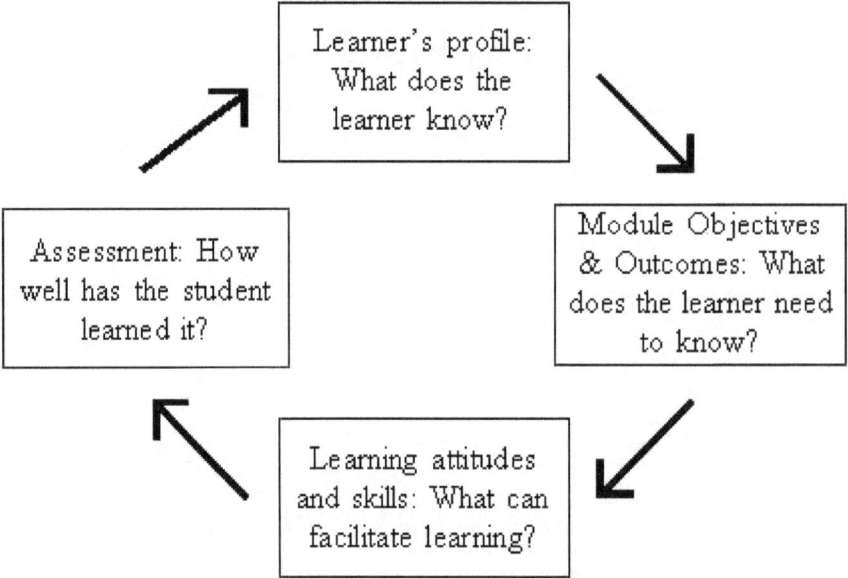

<u>EMMY NOETHER (1882-1935)</u>

Amalie Emmy Noether was a mathematician who is known for her contributions in algebra. Emmy, as she is more often called, was born in Germany and completed her early education at Städtische Höhere Töchterschule (City Highschool for Daughters). In 1907, she became the second women to earn her Ph.D. from the University of Erlanger. After completing her doctorate, Emmy substitute lectured for her father at the University of Erlangen and later lectured at Göttingen College.

Emmy researched several important algebraic ideas, particularly those dealing with noncommutative algebra (a type of algebra where the order of multiplied numbers affects the answer). Emmy's work first attracted attention for a paper she had co-authored with a colleague entitled, *Concerning Moduli in Noncommutative Fields, Particularly in Differential and Difference Terms.* Due to her Jewish heritage, Emmy was dismissed from her teaching post at Göttingen University when the Nazis took power in 1933. The Nazi policy was to remove all Jews from university teaching positions. She then traveled to America where she worked at Bryn Mawr College as a visiting professor and researcher.

DISCUSSION QUESTIONS:

1. What does algebra mean to you?

2. If you had been a professor like Emmy when the Nazis came to power, what would you have done?

GRACE HOPPER (1906-1992)

Admiral Grace Murray Hopper was one of the first female computer scientists and was key in the early development of computer languages. Grace was born in New York and was educated at Hartrige School in Plainfield, New Jersey, where she graduated with a degree in mathematics and physics. She then attended Yale College and graduated with a Masters of Arts and a PH.D in mathematics. After receiving her education, Grace worked as a professor and later joined the Naval Reserve in 1943. Working with the Navy gave her the opportunity to work on a project with Harvard University to create mathematical tables for the navy's Bureau of Ordinance. Grace also worked with the *Mark I* computer project (a project funded by the International Business Machines Cooperation and the U.S. Navy that developed a digital programmable machine that has been said to mark the beginning of the digital computer age).

Grace became head programmer for the Universal Automatic Series of Computers (UNIVAC I), an early line of computers that were sold between 1951 and 1957. She also helped to create computer compilers (a computer program that translates something written in programming language into another computer language) for COBOL, which was the most commonly use programming language in the United States for twenty years. Grace continued to work for the Navy until 1986 and eventually became a commodore as well as the oldest serving officer. She was a public speaker and gave two hundred or more presentations a year, sharing her passion about the fast growing field of computers with audiences everywhere. Grace earned the National Medal of Technology in 1991 and continued to be active in her field until eighteen months before her death. She is recognized as a pioneer in the field of computer technology and has been nicknamed "Amazing Grace."

DISCUSSION QUESTIONS:

1. How do women manage high career goals manageable?

2. What about Grace's story inspires you the most?

EXERCISE #2
WRITE YOUR STORY: ASPIRE!

Begin writing your story by answering these questions:

1. Where and when were you born?

2. What have you studied that you really enjoyed?

3. What do career or careers are you thinking of pursuing?

4. How does that career relate to STEM fields?

5. What will you study to get the degrees you need?

GUILIANA TESORO (1921-2002)

Guiliana Tesoro is an internationally known organic chemist who made significant contributions in the field of textile chemistry. She graduated from Yale University with her Ph.D. in organic chemistry in 1943 and went on to work as a research chemist. Giuliana is best known for her work in the area of polymers. While doing research for textile companies, she contributed to important advancements in flame retardants and polymer flammability.

Guiliana also designed chemical processes to prevent the accumulation of static electricity and to achieve permanent press qualities. She was president of the Fiber Society (society that works to improve knowledge of fibers and fiber-related products) and served on various committees of the National Research Council and the National Academy of Sciences.

DISCUSSION QUESTIONS:

1. What would our clothes be like without the research done by people like Guiliana?

2. What do you imagine will be the chemistry of our clothing of the future?

HATTIE ALEXANDER (1901-1968)

Hattie Alexander is a physician who is internationally recognized for creating an anti-influenza serum that reduced the number of children who died from meningitis. Hattie attended Goucher College in Baltimore, Maryland. She then worked for the U.S. Public Health Service and Maryland public health service as a bacteriologist before continuing on to earn her M.D. from John Hopkins University School of Medicine. While interning at John Hopkins Hospital, Hattie began to take an interest in influenza meningitis and, taking inspiration from previous researchers, she went on to conduct experiments using rabbit serums.

Hattie developed a successful cure for influenzal meningitis in 1939. Consequently, the number of infants dying from the disease dropped to practically nothing. Hattie worked during World War II under the Secretary of War on the influenza commission and served as president of the American Pediatric Society. She continued to research and studied the genetic mutation of bacteria. She published 150 academic papers and served as a professor and lecturer.

DISCUSSION QUESTIONS:

1. How does medical research benefit society?

2. Why would a person want to do research on treatment for a contagious disease?

<u>HELEN NEWTON TURNER (1908-1995)</u>

Helen Newton Turner is an Australian geneticist who headed the team that discovered and studied the FecB gene, the first major gene to have a significant effect on animal production. She attended University of Sydney and, after working in the field of architecture for a short period of time, began working for CSIRO (Commonwealth Scientific and Industrial Research Organization) at McMaster Animal Health Laboratory. Helen later became attracted to the field of statistics, and, with the help of Sir Ian Clunies-Ross, she traveled the UK to study it further.

In 1939, Helen returned to Australia and was reemployed at CSIRO under the Division of Mathematical Statistics as a consulting statistician in the Division of Animal Health and Production. After working this as a statistician, Helen became interested in the application of statistics to sheep breeding. In 1956, she became the leader of the Animal Breeding Section under the Animal Genetics Division in charge of the sheep breeding research at CSIRO. Helen later became more involved in overseas research in places such as Argentina, Lebanon, Peru, Malaysia, Mexico, Pakistan, and Fiji. She was recognized in her field as a world expert, and was awarded the OBE award (Order of the British Empire) in 1977 and the AO award (Order of Australia) in 1987.

DISCUSSION QUESTIONS:

1. Are you interested in pursuing the study of statistics? Why or Why not?

2. Try to name at least three areas of study in which an understanding of statistics is necessary.

HYPATIA (AD 350-370 – A.D. 415)

Hypatia was an early astronomer and philosopher and the first recognized female mathematician. She was a leading mathematician and astronomer during her time as well as a prominent teacher and lecturer. A large part of Hypatia's work related to preserving accurate knowledge for future generations. She was the daughter of an Egyptian astronomer and mathematician, Theon, who is remembered for helping to preserve the *Elucids* (elements) and together they coauthored commentaries on the work of Ptolemy, a Greek astronomer living in Alexandria at the time. After Theon's death, Hypatia continued his work of preserving Greek knowledge of astronomy and mathematics.

Hypatia is also recognized for her commentaries on Diophantus of Alexandria's *Arithmetic*, Apollonius of Perga's *Conics*, and an astronomical table (which some believe may have been a revised version of a part of her father's commentary on the *Almagest*). During her lifetime, she was recognized as the world's leading mathematician and astronomer and was a popular philosophical lecturer and teacher attracting many students and presenting to large audiences.

DISCUSSION QUESTIONS:

1. Why is it important to preserve accurate knowledge for future generations and is this something we should be concerned about this today?

2. What is astronomy and how is mathematics related to the study of astronomy?

JANE COOKE WRIGHT (1919-   )

Jane Cooke Wright is successful physician who is known for her cancer research. Jane was born in New York City to Corinne Cooke and Lewis Wright. Jane's father was one the first African American graduate from Harvard Medical School and Jane followed in his footsteps, graduating from New York Medical College with honors in 1945. After briefly working for the New York City Public Schools, Jane went to work with her father at Harlem Hospital. Together Jane and her father researched anti-cancer chemicals.

After her father's death in 1952, Jane was given the position that previously had been held by her father head of the Cancer Research Foundation. She developed new methods of using chemotherapy and studied different ways of treating cancer. Jane had a very successful forty-year career and published over 75 papers on chemotherapy. She was also the first woman to become president of the New York Cancer Society.

DISCUSSION QUESTIONS:

1. What would you want your children to learn from you?

2. What has inspired you about the medical professionals you know?

JEWEL PLUMMER COBB (1924-   )

Jewel Plummer Cobb is an African American research scientist who is best known for her cancer research. She was born in Chicago, Illinois to Frank, who was a physician, and Carrabelle, who was school teacher. Jewel was third in four generations of medical practitioners, the first being her grandfather, a former slave, who became a pharmacist. She graduated from Talladega College with her bachelor's degree in 1944 and went on to earn her master's degree from New York University and her Ph.D. in cell physiology (a study of how cells work and interact with other cells and the environment around them).

Jewel served as president of California State University at Fullerton from 1981 to 1990. She was the first African American female president of a major west coast university. She was elected a member of the National Science Board and received the Kilby Award for lifetime achievement as well as several other awards. In addition to her research and teaching, Jewel has written about racial and sexual discrimination and raised money to help minorities who were interested in working in her field.

DISCUSSION QUESTIONS:

1. What would you find interesting about teaching STEM?

2. What are some opportunities that are available to help young people who are interested in in a career as a medical professional or researcher?

KADAMBINI GANGULY (1861-1923)

Kadambini Ganguly was India's first modern female physician. She was born in India when it was a British colony. Her family fcame rom what is currently known as Bangladesh. Kadambini's father, the headmaster of a school in the city of Bhagalpur, played a key role in the movement for women's emancipation in Bhagalpur and in the establishment of the first woman's organization in India, Mahila Samiti (established in 1863). Kadambini attended Calcutta Medical School and was awarded a Graduate of Bengal Medicine College degree in 1886. She then travelled to the United Kingdom where she receive a LRCP (Licentiate of the Royal College of Physicians) and a LRCS (Licentiate of the Royal College of Surgeons) before returning to India in 1992. She worked briefly for Lady Dufferin Hospital before setting up her own practice.

In addition to her medical career, Kadambini was a delegate for the Indian National Congress in 1889 and was involved along with her husband in various social causes such as female emancipation, movements to improve working conditions for female coal miners, and the organization of the 1906 women's Conference in Calcutta. Kadambini was one of the British Empire's first two graduates (the other being Chandramukhi Basu) and one of the first women to graduate from Calcutta University.

DISCUSSION QUESTIONS:

1. What do you think might have been some social/cultural challenges that Kadambini faced as one of the first Indian doctors to study western medicine?

2. What characteristics do you think helped Kadambini to succeed?

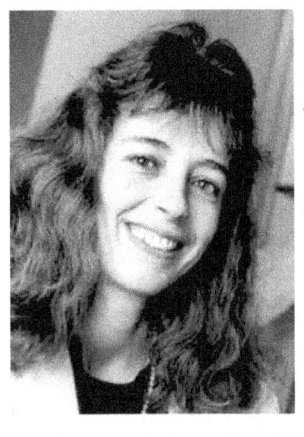

<u>KARIN BLAKEMORE (1953-   )</u>

Karin Blakemore is a physician who has done groundbreaking work in the field of genetics. She graduated with her B.A. degree from the University of Pennsylvania. She then went on to earn her M.D. from the Medical College of Ohio at Toledo in 1978 and a post-doctoral fellowship in human genetics from Yale. After completing her education, Karin worked as the director of the chronic villus sampling program, which was a new process developed for the purpose of diagnosing some birth defects during the first few months of pregnancy.

Karin later worked at John Hopkins School of Medicine where she was on a team responsible for developing a model that worked to provide information which eventually helped with utero transplantation for the purpose of battling genetic disorders. Karin is certified by several important boards such as the American Board of Medical Genetics and has been recommended to the "Best Doctors in America."

DISCUSSION QUESTIONS:

1. What are some of the contributions to healthcare made by medical research in the area of human genetics?

2. If you were to become a physician, what age group would you like to treat?

LILLIAN GILBRETH (1878-1972)

Lilian Gilbreth was a successful psychologist and engineer. She attended the University of California, Berkeley earning a bachelors and a master's degree in literature. Lillian married Frank Gilbreth, a builder and contractor. Together they established a consulting engineering firm, traveled, guest lectured, and coauthored several books including *Motion Study* (1911) and *A Primer of Scientific Management* (1912).

After her husband's death in 1924, Lillian became president of their consulting engineering firm. Unfortunately, many of the factory owners who had been willing to work with her husband cancelled their contracts because they were not interested in taking the advice of a woman. However, Lillian was eventually able to rebuild her husband's business by shifting its concentration and reaching out to more women. She worked as the professor of management for Purdue University in 1935 and in 1921, she was the first woman to become an honorary member of the Society of Industrial Engineers. Lilian continued to do new research, working to develop special equipment for handicapped people. She also continued to lecture and write into her eighties.

DISCUSSION QUESTIONS:

1. What characteristics account for Lillian's long and varied career?

2. What aspects of Lilian's career as an entrepreneur do you find interesting?

MABEL KEETON STAUPERS (1890-1989)

Mabel Keeton Staupers is an American nurse who is known for her work during World War II in the desegregation of the Armed Forces Nurse Corps. Mabel was born on in Barbados and immigrated to the United States with her father. She attended grade school in Harlem, New York and then went to Washington D.C. where she attended Freedmen's Hospital School of nursing. Mabel returned to New York City a few years later and eventually became the executive secretary of the New York Tuberculosis and Health Association and the National Association of Colored Graduate Nurses. During World War II Mabel started a movement to integrate black American nurses into the Armed Forces Nurse Corps. Mabel's campaign was eventually successful, and by January of 1945 the U.S. Army and Navy fully accepted black nurses. She also worked to racially integrate the American Nurses Association, a goal that was reached in 1948.

DISCUSSION QUESTIONS:

1. What do you think is the significance of what Mabel did?

2. What long term impact do you think her work might have?

MARIA AGNESI (1718–1799)

Maria Gaetana Agnesi was a mathematician and philosopher born into a wealthy Italian merchant family. She was tutored at home and showed brilliance from an early age, learning Greek, Latin, and Hebrew. In 1748, Maria wrote *Instituzioni analitiche ad uso della gioventù italiana (*"Analytical Institutions for the Use of Italian Youth"), which was a two-volume publication that covered algebra and calculus in detail. Maria's work helped make popular the use of a cubic curve known as the Agnesi curve, commonly known as "the witch of Agnesi." She was known internationally and her writing was published in French and in English.

Maria was later appointed by Pope Benedict XIV as a professor at the University of Bologna in 1750. She was the first woman in Europe to succeed in establishing herself as a reputed mathematician. Some argue that she was an important component in the Catholic enlightenment in Italy. In her later years, Maria dedicated her time largely to charitable work and religious studies.

DISCUSSION QUESTIONS:

1. What kind of characteristics do you think Maria needed in order to succeed internationally?

2. For what achievements would you like to be known?

MARIE CURIE (1867-1934)

Marie Curie was a French scientist who contributed enormously to the field of physics and is particularly famous for her work in radioactivity. She was born in Poland under the name Maria Salomea Sklodowska and attended the Russian lycée (an international French school). After the completion of her secondary education, Marie worked as a teacher and a governess until she was able to go to Paris to further her education. She finished first in the program for a *licence* of physical science and second in the program for *licence* of mathematical science. She married Pierre Currie in 1895 and together the couple made dynamic contributions to the field of science including the discovery of polonium and radium. Marie earned her doctorate of science in 1903 and was later awarded the Nobel Prize for Physics for the discovery of radioactivity along with her husband.

After her husband's death in 1906, Marie took her husband's place as a professor at Sorbonne, becoming the first woman to teach there. During World War I, Marie worked with her daughter to develop X-radiography. After the war, she and her daughters went to America to meet with President Warren Harding who gifted her with one gram of radium as a recognition of her work. In 1911, Marie won her second Nobel Prize, this time in chemistry. She is the first female Nobel Prize winner and the only woman to win Nobel Prizes in two different fields. Marie's office and laboratory from the Curies Pavilion Radium Institute are preserved in the Curie Museum.

DISCUSSION QUESTIONS:

1. What do you think it would be like to win a Nobel Prize? Two Nobel Prizes?

2. What are some things that you hope to accomplish in your future career?

MARIE MAYNARD DALY (1921-2003)

Marie Maynard Daly was the first African American woman to receive her Ph.D. Marie enjoyed reading from a young age and took inspiration from her father's interest in science. She attended Hunter High School and later Queens College in Flushing, New York, where she graduated in 1942 magna cum laude with a bachelor's degree in Chemistry. Marie completed her doctorate degree at Cornell University in 1947 and went on to teach at Howard University in D.C. for two years. Later, after receiving a grant from the American Cancer Center Society to conduct post-doctoral research,

Marie went to work at Rockefeller Institute in New York. While there she studied various things including the metabolism and composition of cell nuclei for seven years. She later became a professor at Albert Einstein College of Medicine where she worked for about twenty-six years. In addition to her research and teaching, Marie also gave back to the academic community by working to develop programs and scholarship funds to increase the amount of minority students enrolled in medical and graduate school.

DISCUSSION QUESTIONS:

1. What are some ways that you can give back to your academic community by helping your fellow students like Marie helped younger students in her field?

2. What about Marie's story do you find inspiring?

MARY ELLEN AVERY (1927-2011)

Mary Ellen Avery is an internationally recognized physician who became the first female Physician in Chief at Children's Hospital in Boston and the first woman to become Department Chair at Harvard Medical School. Mary originally became interested in medical school as a result of her friendship with Dr. Emily Bacon, a physician who lived in her neighborhood. Mary attended Wheaton College, graduating summa cum laude with a degree in chemistry in 1948, and then went on to attend John Hopkins School of Medicine. After contracting tuberculosis, Mary became particularly interested in the functioning of the lungs. This interest sparked her research concerning infant's lungs who died of RDS. Her research led her to a discovery that played a major role in helping to significantly decrease the amount of infants who died from RDS (falling from 10,000 in 1970 to 1,460 in 1995).

Mary has won various awards for her work such as the Trudeau Medal from the American Lung Foundation, which recognizes someone who has contributed significantly to the treatment and prevention of lung disease, and the National Medal of Science, an award established by the 86th congress which recognizes someone who has done outstanding work in a particular area of science. She also served as the president of the National Academy of Sciences, a leading research organization.

DISCUSSION QUESTIONS:

1. Reflecting on Mary's relationship with her friend Dr. Emily Bacon, what do you feel is the role that mentorship should play in a person's life and academics?

2. What do you think helped Mary overcome her challenges (such as health complications) and succeed?

MARY FAIRFAX SOMERVILLE (1780-1872)

Mary Fairfax Somerville was a British scientist who is believed to be the first female scientist in English history. As was typical for the women of her time, Mary was did not receive a formal education in her early years. She was taught to read by her mother and further educated herself with books from her family's library. In 1816, Mary and her second husband moved to London where she was able to spend time with several respected scientists and mathematicians.

In 1826, she published her first paper entitled *On the Magnetizing Power of the More Refrangible Solar Rays.* She later published several other works including *Physical Geography,* the first geography textbook in English, and *The Connection of the Physical Sciences*, in which she suggested in one of her later editions that the problems surrounding the calculation of the position of Uranus might indicate the existence of another planet. Her speculation inspired others and prompted the discovery of the planet Neptune. Mary is the first known female English scientist and is credited by some as being the first English geographer.

DISCUSSION QUESTIONS:

1. How would you describe the Physical Sciences?

2. If you could hang out with experts in a STEM field, who would you choose and why?

MARY-CLAIRE KING (1946-   )

Mary-Claire King is a geneticist who is particularly known for identifying a genetically inherited gene called BRA1 (a gene increases a person's risk of breast cancer). Mary's interest in mathematics developed from her early attraction to puzzles. After the death of a childhood friend, Mary became interested in biology as well. She decided to pursue genetics in an effort to combine her two interests. Mary received her B.A. in mathematics from Carlton College in Minnesota and a Ph. D. in Genetics from the University of California Berkeley. In addition to her breast cancer research, she has also studied genetic aspects of schizophrenia, hereditary deafness, and ovarian cancer.

Mary has received various awards such as the Clowes Award for Basic Research from the American Association for Cancer Research, which recognizes exceptional achievement in cancer research, and the Gruber Genetics Prize from the Peter and Patricia Gruber Foundation, which is awarded to a scientist who makes an innovative contribution to genetics.

DISCUSSION QUESTIONS:

1. Do you have an interest in doing puzzles like Mary? What kind of puzzles do you like to do?

2. What skills do puzzles give and how they be applied to SYEM studies?

PATRICIA BATH (1942-   )

Patricia Bath is an ophthalmologist who is known for her invention of the laserphaco, a new method for cataract surgery. Patricia's first interest in the medical field was awakened early in life in response to stories she heard of Dr. Albert Schweitzer, a physician who worked for a time in the Congo. Patricia earned her M.D. from Howard University College of Medicine in Washington D.C. and, after interning at Harlem Hospital, completed her ophthalmology fellowship in 1970.

In response to her experiences during her time spent at Harlem Hospital and Columbia Hospital, Patricia conducted a study which established that documented blindness among whites was about half that of blacks, a result that she attributed to the black community's lack of access to proper ophthalmologic care. In an effort to help remedy this situation, Patricia developed Community Ophthalmology, a program designed to provide ophthalmologic care to underserved individuals. This organization has reached many people who otherwise may not have received any treatment for their very serious eye conditions.

Patricia went on to become a professor at UCLA and later became the first female faculty member of the Department of Ophthalmology at the Jules Stein Eye Institute there. Patricia also co-founded the American Institute for the Prevention of Blindness (AIPB) and became its director. She has had a long career lecturing, traveling, performing surgeries, and conducting research. Patricia battled the pressure of sexism and racism throughout her career, but despite the difficulties she never lost sight of her passion for restoring sight to patients.

DISCUSSION QUESTIONS:

1. Discuss a medical need in your community and how you might help meet this need.

2. What challenges would you have in your academic studies if you were blind?

PATSY O'CONNELL SHERMAN (1930-2008)

Patsy O'Connell Sherman is a chemist who, together with one of her colleagues, Sam Smith, invented "Scotchgard", a product for removing stains and spills. Patsy was attracted to a career in science from a relatively young age, and, in 1952, she became the first woman to graduate with a B.S. in chemistry and mathematics from Gustavus Adolphus College in St. Peter, Minnesota. After completing her degree, Patsy was hired by 3M to work on the development of new fuel lines for jets.

As with most women of this time, it was expected that her job would be temporary. Women were expected to quit when they got married and started families. However, Patsy stayed on until her retirement in 1992. It was while working for 3M that Patsy and Sam discovered Scotchgard after Patsy accidentally spilled a chemical on her shoe. That mistake became a famous invention. Patsy eventually came to hold 16 U.S. patents (13 of which were shared with Sam) on new inventions and was given the position of manager of technical development. She was nominated to the Minnesota Inventors Hall of Fame and the National Inventors Hall of Fame.

DISCUSSION QUESTIONS:

1. Try to think of another example of where a simple mistake led to a useful invention or discovery.

2. What characteristics make a person an inventor?

ROSALIND FRANKLIN (1920-1958)

Rosalind Elsie Franklin was a successful British chemist, X-ray crystallographer (a person who studies crystals and their structures by diffracting x-rays), and molecular biologist who played a significant role in the discovery of deoxyribonucleic acid (DNA). As a little girl, Rosalind dreamed of becoming a scientist. She studied physical chemistry at Newnham College in Cambridge, England and later was offered a fellowship to perform research in physical chemistry at Cambridge. In 1942, Rosalind decided to forego her fellowship to research the physical chemistry of carbon and coal for the British Coal Utilization Research Association to further the war effort.

Rosalind later received a doctorate from Cambridge and went to Paris to study X-ray diffraction technology, which led to information on the formation of graphite that proved useful for the coking industry. She later received her doctorate from Cambridge University in 1945 and was offered a fellowship in 1951 to become a DNA researcher on the John T. Randall as Kings College in London. While there, her DNA research led her find proof that clearly established the structure of DNA. Unfortunately, Rosalind's research was copied without her knowledge by fellow researchers Maurice Wilkins and James Watson, who publicized it and received much of the credit, including the Nobel Prize in Medicine just four years after her untimely death. However, she is still remembered as a brilliant scientist who made dynamic contributions to our modern understanding of DNA.

DISCUSSION QUESTIONS:

1. Reflecting on the actions of Wilkins and Watson, what do you think ethics should look like in science?

2. Do you have a dream career like Rosalind did? Explain.

SALLY RIDE (1951-2012)

Sally Kristen Ride was an astronaut and the third woman to go into space. Sally graduated from Stanford University with bachelor's degrees in English and physics. In 1978, while a graduate student, Sally was chosen for the National Aeronautics and Space Administration (NASA). She finished her PhD in astrophysics and started training for NASA. After completing her training and receiving her pilot's license, Sally became a space shuttle mission specialist. She had her first space flight in 1983 aboard the *Challenger,* establishing her title as the first American woman in space.

Sally served on the presidential commissions selected to investigate the tragic *Challenger* explosion in 1986 and the later incident with the *Columbia* in 2003. When Sally resigned from NASA, she accepted a position as physics professor at University of California, San Diego. In addition to teaching, she also founded an organization for middle school girls interested in science called Sally Ride Science. Sally authored or collaborated on various children books about space exploration. Sally was awarded the Presidential Medal of Freedom (the highest civilian award in the United States) after her death.

DISCUSSION QUESTIONS:

8. Would you want to be an astronaut and go into space? Why or Why not?

9. What would you need to study to work for NASA?

SOPHYA KOVALEVSKAYA (1850-1891)

Sofya Kovalevskaya was a Russian mathematician who was also an accomplished writer political activist and public feminist advocate. Although her father objected to her studying mathematics, Sofya smuggled study materials into her room and studied algebra and trigonometry on her own. Her remarkable abilities were noticed by a local professor who begged Sophya's father to allow her to go to St. Petersburg to study.

After completing her studies in St. Petersburg, Sofya wanted to further her education, but the Russian University system would not accept women, and her father would not allow her to travel outside of Russia. In order to solve this problem and give herself the opportunity to travel, Sophya married Vladimir Kovalevsky in 1868. She was then able to study at the University of Heidelberb in Germany and later under the tutelage private mathematics. Sophya was later awarded her doctoral degree by the University of Göttingen with summa cum laude honors. A theory expressed in one of her doctoral papers is often referred to as the Cauchy-Kovalevskaya theorem and gained her recognition within the mathematical community in Europe. Sofya became a professor at the University of Stockholm in 1889 and an editor of the journal *Acta Mathematica.* She was the first woman in Europe to receive a doctorate in mathematics and the first woman to hold a university post in the nineteenth century.

DISCUSSION QUESTIONS:

1. What qualities do you think Sophya have that helped her to be a successful Mathematician?

2. Why should a person be motivated to study and learn new things?

STEPHANIE KWOLEK (1923-2014)

Stephanie Kwolek is an American chemist from New Kingston, Pennsylvania. She earned her bachelor's degree from Carnegie Mellon University in Pittsburg and then worked at Dupont Company as a laboratory chemist in the rayon department. She chose to stay with the company long-term and eventually become a research associate before she retired after forty years of work. Stephanie is famous for her research in polymers particularly aromatic polymers also known as aramids. Aramids are a type of polymer can be turned into strong flame resistant fibers.

Stephanie's research in aramids eventually led her to invent a material that is called poly-paraphenylene terephthalamide (marketed under the name Kevlar in 1971), which has been used to make lightweight bullet proof vests. In 1978, Stephanie was awarded Dupont's ASM's Engineering Materials Achievement Award (an award that recognizes achievements in materials science), and in 1996, President Clinton presented her with the National Medal of Technology (the highest honor for technical achievement in the nation).

DISCUSSION QUESTIONS:

1. What aspects of chemistry do you find most interesting?

2. What inspires you about Stephanie's story?

TEMPLE GRANDIN (1947-   )

Temple Grandin is a scientist and industrial designer who is known for her work as an animal scientist and behaviorist. Temple graduated from Franklin Pierce College in 1970 with a bachelor's degree in psychology and went on to earn a master's degree from Arizona State University and a doctorate from the University of Illinois at Urbana-Champaign. Temple feels that her experience as an autistic person has helped her to understand a strong fear that is present in both animals and autistic people, which she believes was brought about by a hypersensitivity to touch and sound. She has spent her life working to alleviate this problem in both autistic people and animals finding new ways to lessen this anxiety by creating things such as the "squeeze machine," a device that came to be widely used for autistic children and adults.

Temple has authored various books such as *Thinking in Pictures and Other Reports from My Life with Autism,* and *the Autistic Brain: Thinking Across the Spectrum.* Temple's publications were groundbreaking because previously it had been largely thought that there was not a high level of mental activity in the mind of an autistic person and if there was that it was not accessible. Temple gave important insight that better allowed the world to understand both autistic people and animals. She was awarded the double helix medal in 2011.

DISCUSSION QUESTIONS:

1. What are some unique ways in which you think and learn?

2. How might this uniqueness be used to make a powerful difference for good?

VIRGINIA APGAR (1909-1974)

Virginia Apgar is a physician and anesthesiologist who is best known for her contributions to the medical field in the form of the Apgar test. Virginia received her A.B. degree from Mount Holyoke College and her M.D. from Columbia University College of Physicians and Surgeons. She then worked in surgery for two years at Columbia-Presbyterian Medical Center before, specializing in anesthesia. After finishing her studies in anesthesia, Virginia concentrated her research in obstetrical anesthesia (anesthesia for women giving birth). During her time spent dealing in obstetrics, she established a scoring system to assess the health of newborns. Virginia's method was a simple assessment of heart rate, respiration, movement, color, and irritability. The Apgar test is still performed in hospitals today, and it is used to assess how the baby tolerated the birthing process and how she/he is doing outside the mother's womb.

Virginia Apgar was well recognized by a variety of organizations and was inducted into the Women's Hall of Fame in 1995. She was the first woman to become a board certified anesthesiologist and the first woman physician to become a full time professor at the Columbia Presbyterian college of Physicians and Surgeons.

DISCUSSION QUESTIONS:

10. What are something(s) that you are good at or have developed efficient methods for, like Virginia's Apgar test, which you can share to help others?

11. What aspects of medical and health related careers do you find most interesting?

VIVIAN W. PINN (1941-   )

Vivian Pinn was raised in Lynchburg, Virginia and attended the segregated schools there. She wanted to become a physician from an early age and her family encouraged her to work hard to achieve her goals. She attended Wesley College and then went on to earn a M.D. from the University of Virginia, where she was the only African American woman to graduate in her class. She went on to Massachusetts General Hospital to complete her residency and worked as a teaching fellow at Harvard Medical School. She held faculty positions first at Tufts University School of Medicine and later Howard University College of Medicine.

Vivian became chair of the Pathology Department at Howard, making her the first African American woman to hold that position at the school. Vivian received the Elizabeth Blackwell Award from the American Medical Women's Association (an award given to a woman physician who has significantly helped the cause of women in medicine). She was also the first African American woman to serve as the director of the Office of Research on Women's Health at the National Institutes of Health.

DISCUSSION QUESTIONS:

1. Why do you think medical research is an attractive field for women?

2. What are some challenges that Vivian might have faced and what do you think might have helped her to overcome those challenges?

WANGARI MAATHAI (1940-2011)

Wangari Maathai was an environmentalist and political activist from Kenya. Wangari's parents were peasant farmers who provided her with a positive environment in which to grow and dream. After Kenya gained its independence, the country needed educated individuals to fill positions that were previously occupied by the British. Consequently, Wangari had the opportunity to be a part of a group of three hundred students that were flown to the United States to receive a college education. She attended Mount St. Scholastica College and graduated with a B.S. in biology in 1964. Wangari went on to complete a M.S. at the University of Pittsburgh and a PhD at the University of Nairobi. She became a professor at the University of Nairobi and eventually became the chair of the Department of Veterinary Anatomy.

In 1977, Wangari founded the Green Belt Movement (GBM) which implemented a plan of planting trees to slow deforestation. GBM's goal is to involve local village communities in the process. The organization has a vision of a grassroots association of local communities working together to improve their livelihood. GBM developed the Pan African Green Belt Network to educate world leaders about environmental conservation and, over the years, other African Countries such as Zimbabwe and Ethiopia have developed comparable movements. In 2004, Wangari was awarded the Nobel Peace Prize for her work making her the first black African woman to receive this honor.

DISCUSSION QUESTIONS:

1. What are some resources for helping the environment that are available in your school and your community?

2. What do you think scientists should study so that they can make a difference in the environment as Wangari did?

APPENDIX

GROUND BREAKING STORYTELLING

2015 WOMEN in STEM

WOMEN GROUND BREAKERS was founded in 2001 as the Women's Council on Diversity in Chattanooga TN. The group is an ongoing source of cutting-edge programs and resources for diversity, inclusion, and cultural expertise. In 2007, we launched the American Diversity Report using social network technology to educate about cultures along with a Global-Southern Leadership Class. Our materials are used by community and professional leaders in 40 countries. Women Ground Breakers celebrates Women's History by hosting an annual storytelling event and a student video storytelling video project.

Videos of the 2015 STEM Women are available at www.womengroundbreakers.com

Women GroundBreakers 2015 planning committee:

Chair Deborah J. Levine, Earl Berkun, Finn Bille, Sheila Boyington, Jessica Broadnax, Linda Murray Bullard, Natalie Dewhirst, Lisa Diller, Laura Hessler, Gay Morgan Moore, Linda Moss Mines, Carrie Stefaniak, and Sue Stohlmann.

Moderator SHEILA C. BOYINGTON: "What the Trends in STEM mean for Women and Girls."

Sheila Boyington served as Emcee for 2015 Women Groundbreakers Storytelling. She is the President of Thinking Media/Learning Blade® and National Senior Advisor, STEMconnector®/Million Women Mentors®. Sheila has won numerous awards for her Entrepreneurship and Leadership including the Athena, Navigator of Entrepreneurship, Supernova, and Chattanooga Engineer Entrepreneur of the Year. She holds a Masters Degree in Civil/Environmental Engineering from the University of California at Berkeley and a B.S. in Chemical Engineering from the University of Florida.

TRENDS: We need to grow our STEM workforce. STEM jobs are expected to grow 17 % by 2018, but the number of college graduates in STEM fields continues to decline and is down 24 % from two decades ago. Women make up 50 % of the college-educated workforce, but only 14 % of engineers are women and just 27 % are in computer science and math positions. A lack of ethnic diversity means that only 6 % of STEM workers are Hispanic and African American.

EDUCATION & CAREERS: Women are a large potential pool of STEM workers. Yet, college freshmen who express an interest in STEM are 44 % male and only 15 % female. For every 8 boys that plan to pursue STEM, only 1 girl does. Girls often pursue STEM career pathways when they know about them, but they often have no exposure to the possibilities. Girls choose STEM more often when it's taught in the context of helping society and making a difference, but that is often not the case. Creative ways to expose and mentor women and girls can and will lead to an increase in women in STEM.

DISCUSSION QUESTIONS:

1. Why do you think that the number of girls choosing a STEM education is decreasing?

2. Describe how you would inspire your girlfriends to consider a STEM career.

ALYSSA J. MONTAGUE: *"It's a Man's World: The Commercial Construction Industry."*

Alyssa J. Montague is Document Control Manager at Hutton Construction, Inc. She spent "Take Your Daughter to Work Days" at a water treatment plant with her father bonding with him over construction, auto repair, golf, and hockey. Defying the usual roles of women in construction, she uses her IT expertise to build a new filing system, train coworkers, and trouble-shoot her creation.

TRENDS: Smarter automation functionality and ability to customize will be game changers in the future. Technology changes so quickly, that it's the person who keeps up-to-date on the latest and greatest who will excel. Companies are always looking for ways to get more done faster and take advantage of advances in the automation of workflow. They are continually looking for the right person who is willing to go above and beyond to customize and streamline the way things get done.

EDUCATION & CAREERS: The world needs more engineers, scientists, and technology and math specialists to lay the groundwork for change. The news is scary these days, with more violence, bigger natural disasters, global warming, corrupt politicians, war -- the list goes on. Unless we make a concerted effort to educate our youth in the basic STEM fundamentals, humanity is going to continue to blunder its way through life. We're slowly killing our world, and ultimately ourselves, unless we can change the way people think, and quickly. The person with the solution to one of these problems could be out there, but without guidance and education, nothing will change.

DISCUSSION QUESTIONS:

1. Describe how you've gone, or wish you went, above and beyond expectations.

2. What aspects of technology and its ability to make a better world would you like to explore?

 **HEIDI HEFFERLIN:** *"The Passion to Build Things."*

Heidi Hefferlin is an architect and the president at Hefferlin + Kronenberg Architects, PLLC. Her family mixture of bankers, academics, entrepreneurs, ministers, immigrants, and farmers led to a global mindset early in life. Heidi first acquired her love of architecture from her family in Switzerland. In addition to obtaining an architecture degree, she worked in construction to get first-hand knowledge of the trade. Now an entrepreneur, her mission reads, "...founded on the principal that we can improve the lives of those we serve through outstanding design."

TRENDS: In the next decade, buildings will not only be environmentally sustainable, but also adapt to their environment and positively give back. They will generate electricity, rotate to follow the sun, filter and store water. Architecture marries art and science in a wonderful career. The variety and depth of the experience is only limited by your desire to pursue the process.

EDUCATION & CAREERS: There are many avenues in architecture that you can pursue including theory, academia, corporate architecture, and small practice. Having worked in several of these areas, I enjoy the small practice and the actual building of the buildings. The drawings and models refine ideas, but I enjoy most seeing the buildings taking shape and people inhabiting them. My favorite part is experimenting with ideas/designs and building pieces with my own hands. Not surprisingly, our HK mission statement reads, "we are founded on the principal that we can improve the lives of those we serve through outstanding design."

DISCUSSION QUESTIONS:

1. What skills do you have that could be useful as an architect?

2. What aspects of this career do you think you would enjoy?

JEMILA MORSON: *"The Power of We."*

Jemila Morson is Owner and Creative Strategist at MorSocial Media. A blend of cultures, Jemila calls herself "the perfect mix: an Indian, American, Caribbean chick." Jemila is an entrepreneur who uses her skill in digital media management to assist small businesses to brand and market themselves. She created the MorLadies online networking group to empower and inspire other women business owners.

TRENDS: Socially intelligent and interactive technology is our exciting future. Tools already exist that combine group intelligence to connect or create augmented realities. Think of wearable smart devices such as google glasses or smart watches. Another program setting the tone for the future is called "Aurasma." The free program merges the physical and digital worlds by allowing us to use smart devices to scan objects and engage in video and audio content from a cloud based system. The possibilities are endless and many businesses are already making use of its power.

EDUCATION & CAREERS: Study the field of social technology, not because it's the future, but because it's already here. Every day, new developments lessen the gap between ideas, industries, and people. Though still a new and fragile field, it's become an essential part of doing business. Young people should be building the future of social technology or, at the very least, learning how to market and leverage it in whatever field they choose to pursue.

DISCUSSION QUESTIONS:

1. What about Jemila's story reminds you of your own story?

2. What aspects of Jemila's career do you find intriguing?

 **LAKWESHIA EWING:** *"LIFE APPlications: Owning your Legacy."*
Lakweshia Ewing is a co-owner at Biz Boom Apps, LLC. She was born into poverty, but dreamed of becoming a game-changing, breakthrough pioneer of new technology. She has degrees in Psychology, Educational Administration, and is a doctoral candidate in Organizational Leadership. Lakweshia has a passion for service, whether through her youth ministry, helping small business owners, or serving on the board of numerous civic groups.

TRENDS: The internet was created so that we have access to any information, from anyone, about anything. We are quite literally drowning in data. While we have created some very useful search engines like Google, even they are having a hard time separating the meaningful information from the meaningless. As a result, over the next decade we will see significant changes in how we interact with the internet. We're already seeing those changes with websites like Wolfram Alpha which "computes" answers to queries rather than simply returning search hits, and Microsoft's Bing, which helps take some of the guesswork out of searches. As technology and devices like phones, TVs, computers, and cars become increasingly connected, we should prepare for rapid changes in how we interact with and make sense of the internet.

EDUCATION & CAREERS: As the workforce evolves, STEM knowledge and skills are becoming more necessary in many professional arenas. We face both the challenges and successes of a knowledge-based global economy. Technological and scientific innovations are the future of our society's growth. Yesterday's STEM strategies will not sustain students in this new information age. Young ladies of today must develop their educational capacities to higher levels, beyond the innovations of the past.

DISCUSSION QUESTIONS:

1. What kind of information on the internet is meaningful to you?

2. What kinds of technology and devices would you enjoy working with as a career and why?

 **DR NESLIHAN ALP: "The Vision of an International Woman in STEM Education."**

Dr. Neslihan Alp is Interim Dean at UTC College of Engineering and Computer Sciences. Born in Turkey, Dr. Alp learned French before learning English and never saw a computer before emigrating for an engineering degree. She became an online learning pioneer, a mother of two, and one of UTC's youngest faculty members. She progressed to Department Head, Assistant Dean, and is the only woman to become Interim Dean.

TRENDS: An exciting development in STEM is a major trend in the corporate world that emphasizes team work. Employers now need not only STEM skills, but also soft skills that can be applied to team building and leadership development. In addition, there is a need for cross-over expertise and for the flexibility to do multi- tasking. Management positions will require these skills at all levels, on the manufacturing floor, with vendors, partners, clients, and community leaders.

EDUCATION & CAREERS: STEM (Science, Technology, Engineering, Math) fields are our future, and young people should be open to the opportunities that STEM education can bring them. Young women in particular should follow the emerging trends closely. They can combine a STEM education with the decision-making, communication, and sensitivity skills that are so often nurtured in women since childhood. Women can place data into a human context, anticipate the audience's response, and customize their approach for better results. Women can combine their people skills with technical skills and increase their future career opportunities.

DISCUSSION QUESTIONS:

1. What STEM field appeals to you and why?

2. What skills do you have that could be applied to a management position?

SONYA REID: "My Non-traditional STEM Journey."

Sonya Reid is a Business Process Improvement Manager at Erlanger Health System. Determined to pursue her dream of a college education even with responsibilities for five younger siblings, Sonya was inspired by teachers and a Career Fair to get her degree in Chemical Engineering. After working in the field, marrying, and having children, Sonya obtained a Masters degree in Engineering Management. She now applies her expertise to healthcare, improving hospital management on behalf of the professionals and the patients.

TRENDS: I'm excited that engineering management principles are being applied to healthcare in many hospitals, adding to the quality and efficiency of patient care. I anticipate that this trend will continue. There will be new career opportunities for individuals with process improvement expertise who traditionally have worked in manufacturing environments.

EDUCATION & CAREERS: Today, there are many applications for STEM and young people should be introduced to the opportunities in various fields early in their education. I hope that young people have mentors in their teens, as I did, who can coach them if they have an interest in a STEM field. However, I do encourage students to pursue their passion, not just a potential job prospect. Both my daughters are interested in STEM fields and I have encouraged that interest. My oldest is a sophomore Biomedical Engineering student, and my youngest, a high school senior, was recently accepted into college and into their Computer Science program.

DISCUSSION QUESTIONS:

1. What kind of person would you choose to mentor you?

2. If you were an expert in engineering management, where would you apply your organizational skills and or whom?

 **TAMRA VAN ALLEN: "Applying My Math Skills."**

Tamra Van Allen is Assistant Vice President and Pricing Actuary at Unum. She grew up in a conservative religious school where her female cohorts focused on and competed in STEM subjects. Inspired by her grandmother who pursued STEM in the 1930s, Tamra spent lonely years studying for and successfully taking the actuarial exams. She now integrates mathematical skills, design, and behavioral technology to help people get sustainable health insurance.

TRENDS: The exciting development in my field involves behavioral technology. Integrating design and the consumer experience into mathematical skills is where the future is headed. The impatience that the "Amazon phenomenon" of the retail arena has built into the modern world is bleeding over into healthcare and purchasing insurance. People are impatient with a process that takes too long, so actuaries have to figure out how to get the information they need more quickly to the consumers, while making sure their analysis is stable for long term. The next generation of insurance will be able to translate all the formulas for risk into something that is user friendly.

EDUCATION & CAREERS: There are many reasons why a young person should consider my field. There is great job security, partly not everyone has the discipline it takes to finish the actuarial process. It is also a great education because the process is so challenging and we have many opportunities to be a leader and shape the field. Women fill the newer fields much more heavily compared to the old-school insurance arenas. There is a huge demand for men and women who like design and behavioral science and can integrate that with their math skills.

DISCUSSION QUESTIONS:

3. What do you like about math and what do you dislike?

2. How would you use math to help others and make a difference?

 **DR. RUTH WILLIAMS:** *"At Least She Was Never Bored."*

Dr. Ruth Williams is a PsychoBiology Professor at Southern Adventist University. A blend of Black Caribbean, Dutch, French, and British cultures, Ruth immigrated to America from the West Indies where a degree in psychology was nonexistent. Despite the family's hardships, her optimism and dedication led to a fulfilling career in education. She leads, trains, and inspires students to do research projects that create knowledge through the scientific process.

TRENDS: Advances in technology and science that provided us with the means of neural imaging have broadened our horizons in the understanding of how to explain the why and how of behavior at the level of neuroanatomy and physiology. To actually see what is happening in the brain of a person solving a differential equation, telling her child that she loves him, or reading a sad story, portends not only the ability to describe the biology of behavior but also be able to begin to change some of the dynamics for the greater good.

EDUCATION & CAREERS: Young people today should study the field of educational, neuro, and quantitative psychology because it is on the cusp and cutting edge of advancements in the field of neuroscience. From understanding how children learn ratios and proportions to "rewiring" the neural connections in a veteran returning with Post Traumatic Stress syndrome, the sky is the limit for what my field has to offer.

DISCUSSION QUESTIONS:

1. What questions about the human brain interest you?

2. What character traits account for Dr. Williams' success and which of them do you think you have, too

Sources

IT professionals (IEEE computer society) 17, Issue 1 (January 2015): 62-64.

IT Professionals (IEEE COmputer Society) 17 (2015): 62-64. August 2002. http://www.invention-ifia.ch.

Adkinson, Asiye Yilmaz, and Robert Adkinson. "The FecB (Booroola) gene & implications for the Turkish sheep industry." *Turkish Journal of Veterinary & Animal Sciences*, Dec. 2013: 621-624.

Administration, National Aeronautics and Space. n.d. http://www.jsc.nasa.gov.

Advance Materials & Processes. "In Memoriam." August 2014.

Advance Materials & Processes. "Kwolek presented with National Medal of Technology by..." 1996.

Advancement of Animal Breeding ad Genetics. n.d. http://www.aaabg.org.

African American Heritage Program. http://www.cpnas.org.

African American Registry. http://www.aaregistry.org.

American Chemical Society. http://www.acs.org.

American National Biography. http://www.anb.org.

American Society of Human Genetics. http://www.ashg.org.

Anita Borg Institute. http://anitaborg.org.

Association for Women in Computing. n.d. http://awc-hq.org.

Brown, David. *Inventing Modern America : From the Microwave to the Mouse.* Cambridge, Massachussetts: MIT Press, 2002.

Cal Poly Pomona. n.d. https://www.cpp.edu.

California State University. http://www.calstatela.edu.

Chemical Heritage Foundation. http://www.chemheritage.org.

Cold Srping Harbor Laboratory. http://www.cshl.edu/DHMD/History.html.

Das, Sakti. "2007 History of Urology Forum." *Auanews*, March 2007: 15-16.

Encyclopedia Britannica. http://www.britannica.com/.

Kempis, M. Thomas a. "An Appreciation of Sophie Germain." *National Mathmatics Magazine* (Mathematics Association of America) 14 (November 1939): 81-91.

Kennett, Jeanette. "Autism, Empathy and Moral Agency." *The philosophical Quarterly* (Oxford University Press) 52 (July 2002): 340-357.

Kenschaft, Patricai C. "Charlotte Angas Scott." *The College Mathematics Journal* (Mathematical Association of America) 18 (March 1987): 98-110.

Kessler, Grant, and Jenniffer Judson. "African American Scientists and Inventors." *Black History Bulletin* 70 (2007): 7-11.

Langevin-Joliot, H. "Radium, Marie Curie and Modern Science." *Radiation Research* (Radiation Research Society), November 1998: S3-S8.

Lemelson-MIT. http://lemelson.mit.edu.

Los Angees Times. n.d. http://articles.latimes.com.

Luthra, Inderpal Rajesh. "Shanti Swarup Bhatnagar Prize: women finally show their might." *Current Science*, November 25, 2010.

"Mary Somerville's extraordinary life." *Lancet*, April 2002.

Mazzotti, Massimo. "Maria Gaetana Agnesi: Mathematics and the Making of the Catholic Enlightenment." *Isis* (University of Chicago Press) 92 (December 2001): 657-683.

Mckee, Maggie. "Sally Ride (1951-2012)." *Nature*, February 2015.

Minnesota Science and Technology Hall of Fame. n.d. http://www.msthalloffame.org.

National Inventors Hall of Fame. http://invent.org.

National Library of Medicine. Nationa Institute of Health. n.d. http://www.nlm.nih.gov.

National Science Foundation. http://www.nsf.gov.

Nimbkar, Chanda. "Genetis improvement toincrease sheep meat production and profitability of sheep rearing." *Current Science*, June 25, 2009: 1566-1567.

Pakrashi, S.C. "Asima Chatterjee (1917-2006)." *Current Science*, May 10, 2007.

Raman, Ramya Anantanarayanan Raman. "Surgeon Senjee Pulney Andy's trials in treating smallpox using leaves of Azadirachta indica in southern India in the 1860s." *Current Science*, June 25, 2013.

Rapoport, Sarah. "Rosalind Franklin: Unsung Hero of the DNA Revolution." *The History Teacher* (Society for History Education), November 2002: 116-127.

Rappaport, Karen D. "S. Kovalevsky: A Mathematical Lesson." *The American Mathematical Monthly* (Mathematical Association of America), October 1981: 564-574.

Sanderson, Marie, and Mary Somerville. "Mary Somerville: Her Work in Physical Geography." *Geographical Review* (American Geographical Society) 64 (July 1974): 410-420.

Scholar Space. University of Hawaii. http://scholarspace.manoa.hawaii.edu.

Seneta, E. "In Memoriam: Emeritus Professor Henry Oliver Lancaster." *Australian & New Zealand Journal of Statistics* 44 (December 2002): 385-400.

Silverberg, Alice, and Dirk Huylebrouck. "Emmy Noether in Erlangen." *Mathematical Intelligencer*, 2001.

University of Hawaii. http://www.hawaii.edu.

University at Buffalo College of Arts and Sciences. n.d. http://www.math.buffalo.edu.

Wayne, Tiffany K. *American Women of Science Since 1900: Essays A-H Vol. 1.* Greenwood Publishing Group, 2011.

Wolf, Cary. "Learning from Temple Grandin, or Animal Studies, Diability Studies, and Who Comes After the Subject." *New Formations*, 2008.

"Women Who Nurse: Mabel K. Staupers, R.N." *Rn* 4 (January 1941)

www.ingramcontent.com/pod-product-compliance
Lightning Source LLC
Chambersburg PA
CBHW080549190526
45169CB00007B/2703